a touch of Sweden

LAND SKALL MED
LAG BYGGAS.
SVEAR ÄGA
KONUNG TAGA,
SÅ OCK VRÄKA.

Land shall
be built by law.
The Swedes have a right
to accept and
to reject a king.

This is a wellknown quote from Sweden's oldest law, the Elder Västgöta Law. It was written down in the early 1200s, presumably by lawman Eskil Magnusson, brother of Birger Jarl, who was instrumental in establishing a strong central government and national legislation. But the law's origins are believed to be several centuries older.

This imposing grave on the plains of Västergötland Province stands as a gateway into Swedish history. It was erected over the kinsmen of our farming forefathers 4,000 years ago, during the Stone Age.
Lapporten, the gateway to Sweden, far to the north near the Norwegian border, through which Lapps pass on their annual migrations.
Exceeding 6,000 meters in length, the Öland Bridge, connecting Öland with the mainland, is Europe's longest bridge.
Three thousand years ago the Bronze Age inhabitants of Bohuslän carved magnificent ships on flat rocks in hopes of communicating with the gods. Today's residents must content themselves with communicating with their fellow man through the atmosphere. The circular antenna in Tanum receives TV signals via satellites and can handle 1,200 simultaneous telephone conversations with the United States.
Three quarters of a million persons have immigrated to Sweden during the post-war period. While outwardly different, they are basically similar.
Kalmar Castle was built in the 1100s but was remodelled in the 1500s by King Gustav Vasa and his sons.
Four hundred years later, the art of construction had found broader horizons, as shown by the Essingeleden highway bisecting Stockholm.

POLIS

a touch of Sweden

KW GULLERS

TEXT:
GUNNAR ARVIDSON

PHOTO:
BJÖRN ENSTRÖM~PETER GULLERS

DESIGN:
JAN BOHMAN

GULLERS INTERNATIONAL

THE SWEDES AND THEIR KINGS

It is probably safe to say that Sweden has had kings for close to 1,000 years. However, the country did not become a united kingdom until the nobleman Gustav Vasa placed himself on the throne by force. By the time he died in 1560, he had succeeded in laying the groundwork for the protestant state church and had established an administration, national finances and a functioning nation. His statue impresses even today.

Others carved their name in history with a sword – as kings were wont to do. Gustav II Adolf, who was killed in the Thirty Years' War, Karl X Gustav, who annexed the provinces of Skåne, Blekinge and Halland from Denmark, Karl XII, who drained the country's resources in his military campaigns far into Russia. The popularity of kings in bygone times should not be exaggerated.

This popularity appears to have increased as royal power has diminished. The present king, Carl XVI Gustaf – descended from one of Napoleon's field marshals who was offered the Swedish throne by a single Swedish lieutenant, an unusually successful instance of private initiative – has no other function than to represent his country at official events. One such is the annual Nobel Prize festivities at which the king presents each of the winners with his award.

American author Saul Bellow, winner of the Nobel Prize for Literature in 1976, receives his award from the hands of King Carl XVI Gustaf.

Gustav Vasa belonged to a period of great royal power, symbolized in this giant likeness in oak by Carl Milles. It is found in the Nordic Museum.

WARER
VENSKE

ART IN SOCIETY

Visitors to Sweden may be duped into thinking that this country contains an unusual number of people who are interested in art. This is a pleasant misunderstanding that we will allow to survive.

Mainly during the past few decades, public art has undergone a significant expansion. Lavish reliefs and paintings hang in staircases and vestibules. Hospitals and schools are generally richly decorated. In otherwise gray, monotonous suburbs, areas suddenly sparkle with all sorts of modern art. In subway stations, artists have been given a free hand and large walls on which to use them. Communities strive to outdo one another with public art. This is sensible competition.

New playgrounds no longer consist of three swings and a sandbox as dreary props for 30 little illiterates bopping each other on the head with pails and shovels. There are now more and more dragons or other imaginary figures in gaily colored plastic on which kids can climb. Many a young tot's first encounter with art comes in such a playground and he learns that art needn't be taken so seriously. Art doesn't have to mean kings or sloppily clad goddesses in pigeon-besmirched bronze.

There is an old rule and an even older law that one per cent of the total value of a new public building must be devoted to ornamentation. Art is the first thing to be skimped on when times are tough, but public art continues to thrive and artists receive much-needed support. Those who like and appreciate art can thereby see it in their daily surroundings. Those who don't like cannot escape it.

And art has a purpose, even if just to aggravate.

A child's first provocative contact with modern art could well be playground sculpture such as this one in Stockholm.

TENSTA
TENSTAPLAN
HAGSTRAKET
BUSS 538
barn är allas barn

The lurking tristesse of a modern suburban center can be dispelled by unexpected art, such as here in Flemingsberg outside Stockholm.

In many of Stockholm's subway stations, which otherwise would have been dominated by advertising and graffiti, artists have been permitted to decorate at will.

A break in the monotony of factory routine relaxes the body and refreshes the mind.

Many strollers in Stockholm have been startled by this statue by K.G. Bejemark.

The City Hall in Landskrona, a public building, yet at the same time a work of modern art, designed by architect Sten Samuelsson.

STADS

149

The dream of summer for city dwellers.

THE CITY AND THE COUNTRY

Swedes have never gotten over their dream of a little cottage. Preferably it should be located in the middle of town within easy walking distance from the office. Unfortunately, city planners can't quite meet even this elementary a demand. Instead, Swedes buy every little dilapidated cabin in the country at exorbitant prices. Then they sit in their modern and comfortable concrete suburbs and yearn to get away.

Official policy since World War II has been that everyone has the right to a dwelling with adequate space. This has largely been provided—with the result that many cities suddenly look very much alike. The same shopping center, the same parking lots, the same pedestrian streets, the same white or gray suburbs. They are drab, sterile and rather isolated. But the standard is remarkably high—as are the rents.

Nowadays, when all the little red cottages have been purchased, most people are trying to combine the old dream and the new comfort by building their own small houses. The day of the high-rise apartment building is gone. Bungalow areas are growing in ever-widening circles. Traveling through Sweden and seeing the endless rows of tidy houses and gardens will convince even the most skeptical observer that this is a pretty affluent country.

Now that the urgent postwar building frenzy has subsided, older buildings are coming into their own. No longer are city centers being razed at any cost in order to make room for massive new buildings. It is now *in* to live in an old building, and condemned houses are being restored at great cost. Few people can afford to live in them but there they stand, adding to the beauty of the city.

Poets speak of Stockholm as "The City that Floats on the Water."

The old has found renewed life along Stockholm's Hornsgatan.

Row houses, such as these in Sala, provide a way for more and more people to live in their own homes.

adidas
HP

Agility, grace, harmony—all synonymous with the Idla Girls.
(preceding pages)

An increasing number of Swedes are making a habit of some form of regular exercise. For some it is a pleasure. For others, a struggle.

ABBA – initials now recognized from Moscow to Sydney. Benny Andersson, Anni-Frid Lyngstad, Björn Ulvaeus and Agnetha Fältskog.

The voice of Birgit Nilsson is known and loved by opera lovers the world over.

The choral Society, Orphei Drängar, one of the many fine choral groups carrying on Sweden's tradition of singing for the sheer joy of it.

For many years Helsingborg had Sweden's oldest city theater. When it was razed and replaced by a parking lot, the city got, instead, the country's most modern new theater, directed by Claes Sylwander. The size and shape of the stage can be varied according to dramatic demands and orchestral size. In the control room, the most modern technical, electronic and computer equipment regulate sound and lighting. The theater, decorated in warm red velvet, has 599 seats.

Vad gör Juni?
Vad gör Juni
Jouni sitter
Jouni sitter
på en stol.
Jouni sitter

SENSORISKA MINNET
KORTIDS-MINNET
LÅNGTIDS MINNET

THE SCHOOLS

In 1950 the Swedish school system underwent a fundamental change. This was when the cumpulsory nine-year basic school was introduced. This school was to be open to all, regardless of geographic, social or economic obstacles. This was a major step towards social equality.

Since then the school debate has continued uninterrupted and there has been continued experimentation. The last controversy involves grading. Should grades be continued at all, as a carry-over from an old authoritative school system?

Now the time is ripe for the next significant reform. It is called SIA (Skolans inre arbete), or the school's inner function, and is a sweeping change that seeks to remove the teacher from his desk and give pupils a greater possibility to decide for themselves during classes, to provide things to do at school during pupils' free time, and to promote pupil contact with the home and society.

Parents who barely have had time to familiarize themselves with the old system are now confused and are grumbling in much the same manner as their parents did before them: "I wonder if things weren't better before—in the old school."

Without doubt the reforms have resulted in freer, more outgoing youngsters with a greater understanding of life outside the school. One might say that pupils have a poorer knowledge of many basic subjects.

It's a new and confusing world, especially for the many immigrant children who came from countries with a different and presumably much stricter school system. Immigrants make up seven per cent of the Swedish population. So many may seem bewildered by today's Swedish schools.

Just like many Swedes.

The last day of school—always a joyful one—in Vaxholm, north of Stockholm.
(preceding pages)

Sweden's reformed school system is no longer dominated by the teacher. Pupils are assuming increased responsibility. But the drudgery of learning to spell remains the same, regardless of system.

Celebrating Walpurgis Night in Uppsala, carrying on an old romantic university tradition.

Young people are getting a better education at the same time as older people and many immigrants are finding that today's world places demands beyond their abilities. Adult education is becoming increasingly more common, and employers help by allowing employees to study on company time. Schools also provide baby-sitting assistance where needed.

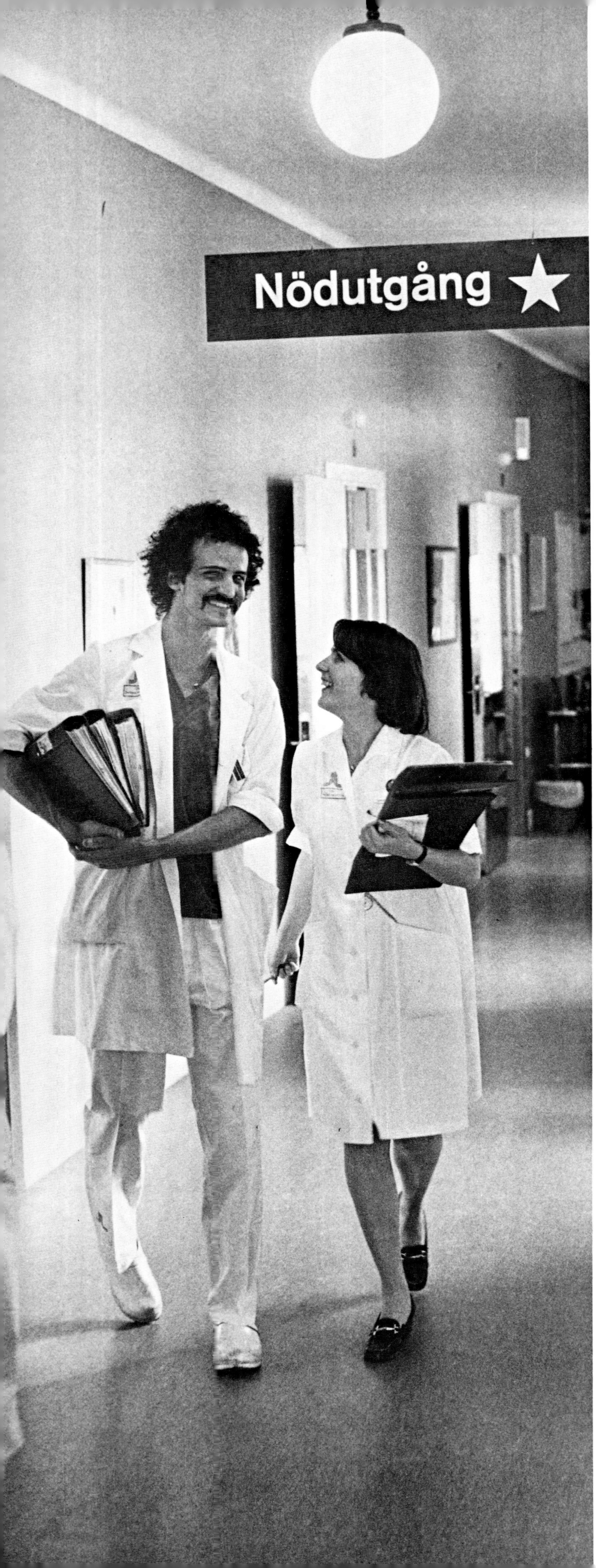

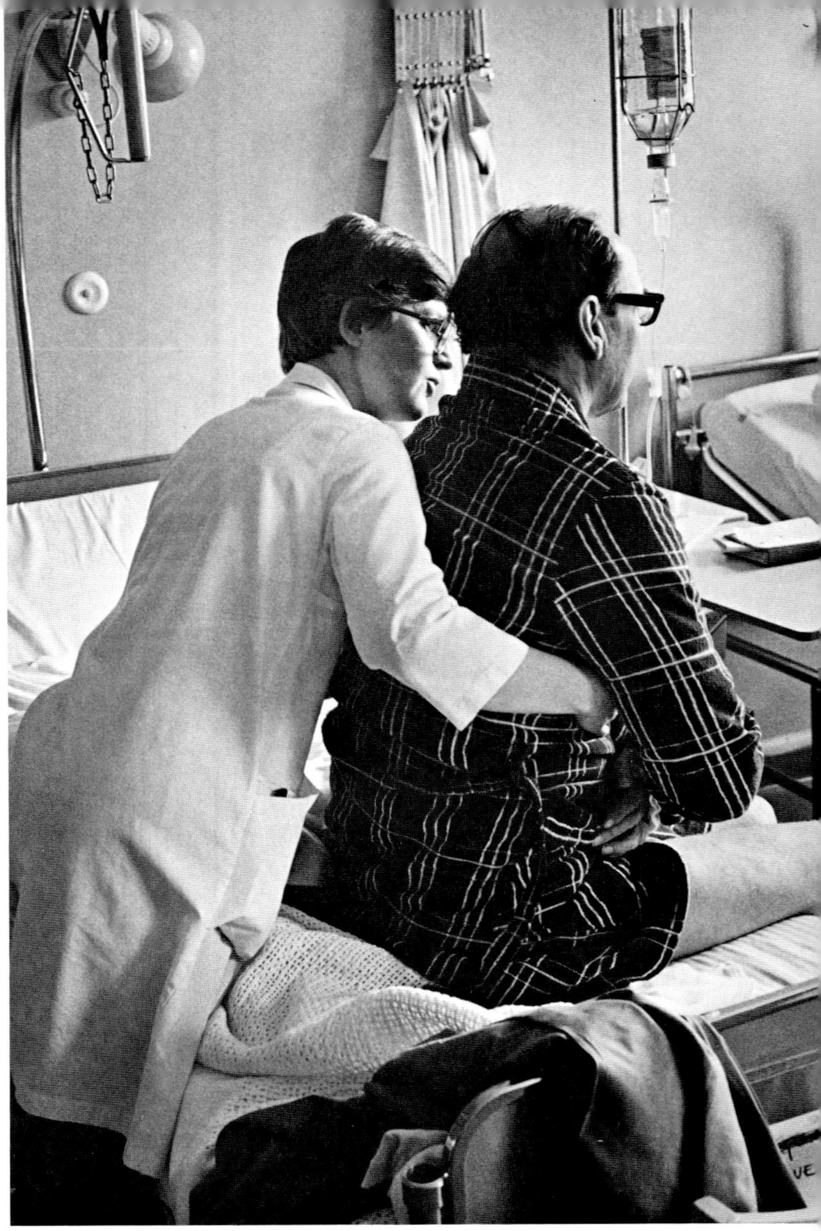

Medical services—where resources permit—are characterized by a spirit of trust and cooperation on the part of the personnel, and the possibility to relax in friendly surroundings.

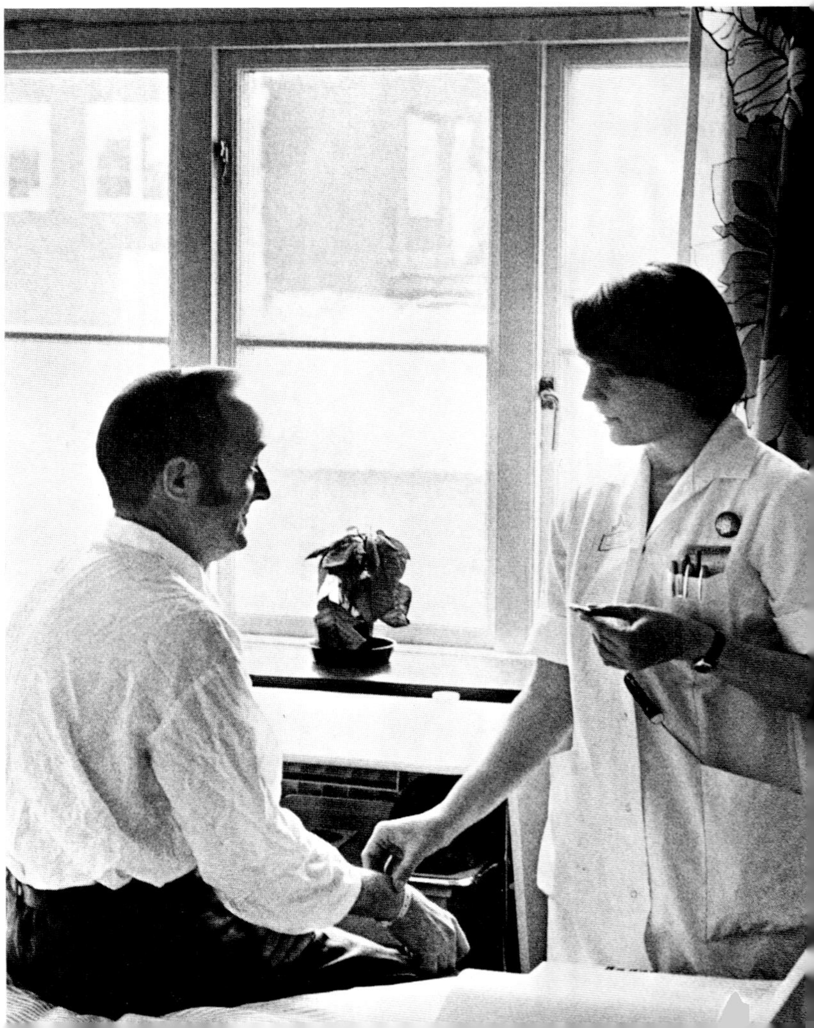

MEDICAL SERVICES

Medical services have developed more rapidly perhaps than any other sector of the Swedish society. In just two decades oedical personnel has tripled and medical care's share of the gross national product has more than doubled. Now, about one in ten gainfully employed Swedes is in some way involved in medical services.

Progress has been too rapid, and there are many reasons for this. The health-care system has been forced to such a degree of effectiveness that we often forget about the care of the individual patient. Both patients and medical staff sense this and are unhappy about it. The system is getting too big and impersonal.

There is no doubt that the technical level of Swedish health care is very high. The new modern ospitals compare well with those abroad. It's the general health care that is carrying almost the entire burden. Less than ten per cent of Swedish doctors have a private practice.

If Swedish health care has grown rapidly, then the demands placed on it have grown even faster. With the general health insurance, everybody can afford to be sick. Everybody can see a doctor, be admitted to a hospital and receive the most expensive medicines at a nominal cost. This is the result of a great reform in the name of social equality, which, however, has added to the strains on health services. There are still too few doctors and nurses to withstand this strain. And all the time the need in sectors like geriatric care increases year by year as life-expectancy rises.

Meanwhile, medical services are in a difficult and costly dilemma and it will take quite a few years to get out of it. In the interim, some people will suffer. This is the other side of the effective welfare society.

ARBRÅ

The Hälsinge Hambo in Järvsö, a gay tradition for those in good shape.

During Walpurgis Night, bonfires crackle throughout Sweden (such as here at Skansen in Stockholm) as a distress signal to the spring from a frozen people.

TEXTILES

As in Europe in general, Sweden's development from a quiet agricultural country to a modern and effective industrial nation began in the late 1800s and proceeded very rapidly. A massive migration from the countryside to the cities often involved abrupt relocations and drastic alterations of life styles. In such circumstances it would have been easy to lose sight of old cultural heritages.

Fortunately, there were people who, even then, understood how to preserve the old arts and skills and pass them on to succeeding generations. There were things like family traditions, folk music, the art of narration, handicrafts and, perhaps above all, the textile art, probably the most broadly entrenched and most vital folk skill.

Even more than 100 years ago, foresighted women established a society with the fitting title Friends of Handiwork in order to preserve and improve "feminine handicrafts." It is still active and continually trains new textile designers. The picture on this page is from the society's premises on Djurgården in Stockholm.

Throughout Sweden, evening courses are full of young women – and some men – who are fascinated by the old art od crocheting and embroidering. This is the revenge of an old craft once in danger of being swept away by a tide of sewing machines. While the Swedish textile industry is in dire straits, home looms are thumping throughout the country.

THE SPIRIT OF SALTSJÖBADEN

The spirit of Saltsjöbaden is a revered concept in the Swedish society, even if it has become a little tarnished and diluted during recent years. It is the symbol of mutual understanding and peace on the labor market.

The name came from Saltsjöbaden, a suburban community outside Stockholm. It was there, in 1938, that employers and trade unions arrived at an agreement that laid the groundwork for the important peaceful labor market following World War II.

Per Albin Hansson was then the Social Democratic Prime Minister. The spirit of Saltsjöbaden also became the necessary prerequisite for the Swedish welfare state, which began to grow under Hansson's leadership.

Marcus Wallenberg, powerful leader of Sweden's largest financial institution and industrial empire for many decades, is in many ways Hansson's opposite. But both men are part of the spirit of Saltsjöbaden, the willingness to cooperate and compromise that led Sweden along the Middle Way.

Per Albin Hansson, Prime Minister from 1932–1946, laid the groundwork for Swedish social welfare. Marcus Wallenberg is the outstanding champion of Swedish industry and finance.

Consultation and codetermination are now legislated realities on the Swedish labor scene. In many companies, such as here at Siemens Elema, manufacturer of specialized medical products, the reform evolved as a voluntary and practical manner of operating.

DEMOCRACY ON THE WORK SITE

MBL looks like the three letters on a Swedish license plate. But they are far more important than that, even in this heavily car-oriented country.

MBL stands for *Medbestämmandelagen,* which is the Swedish name for the recently introduced legislation that provides for codetermination in the realm of business and industry. The law represents a revolution on the labor market–giving employees the right to participate in the decision-making process during all important deliberations affecting the individual work site and work conditions. The old law, which lasted for 70 years, gave the employer the right to direct and allot work operations, and to hire and lay off personnel, has been abolished.

Under the new law, consideration is taken of every employee's experience and knowledge. Employers are consulted when it comes to hiring, firing and transferring, during company cutbacks and expansions, and when a company is sold or appoints a new manager or director.

The transition from one law to the other is understandably difficult. It is hard for employers to see previously held power slip from their grips. And it is tough, too, for employees to assume the added responsibility and demands placed on them.

The employee's safety representative has an increasingly important function to fill, besides her daily work in making components for pacemakers. Siemens Elema produced the world's first cardiac pacemaker.

MINING

When Swedes speak about mines they automatically think about Lapland way up in the north and tend to overlook the belt of mines to be found across central Sweden. Normally, about 95 % of the iron ore that Sweden exports comes from the LKAB-operated mines in Kiruna and Malmberget above the Arctic Circle. Kiruna has the world's largest underground mine, up to 500 meters in depth.

This in itself is nothing to brag about, but the advanced mining techniques applied in the mine are a source of considerable pride. The deep mine shafts provide part of the explanation for the difficulties plaguing Swedish iron-ore mines during recent years. It has been hard to compete with ore obtained at appreciably lower expense from open-pit mines such as those in Australia and Brazil.

Whereas iron ore remains the backbone of the Swedish mining industry, even in adverse times, there are other available ores in Sweden as well. Boliden, for example, has Europe's largest lead mine. Copper, once the best source of national income when Sweden's warring kings needed fast money, is still to be found.

As far as the future is concerned, discussions are underway as to the feasibility of mining the vast shale layers of south-central Sweden to extract their uranium deposits.

STEEL

The forest and steel are the very life stream of Swedish industry. And, admittedly, it has flowed a little sluggishly of late.

One of the first things a Swedish child learns is that "Swedish steel has bite." Indeed, it has "bitten" well for many decades as one of the country's greatest industries. Sweden has found use for all the iron ore it could mine, and Swedish commercial and specialty steels have had a sure and expansive market. The life stream has functioned well without any danger signals. Swedish steel has meant quality.

It still does – but the over-all picture has changed. Rising foreign competition, coupled with a diminished demand on the world market, has caused problems in the steel industry.

Björneborgs Bruk, Björneberg.

Shaping glass in the age-old manner at the Orrefors Glassworks.

An increasingly common view in Swedish industry – industrial robots on the factory floor. Robots like these at ASEA are effective and undisturbed by noise, stress and crowding.

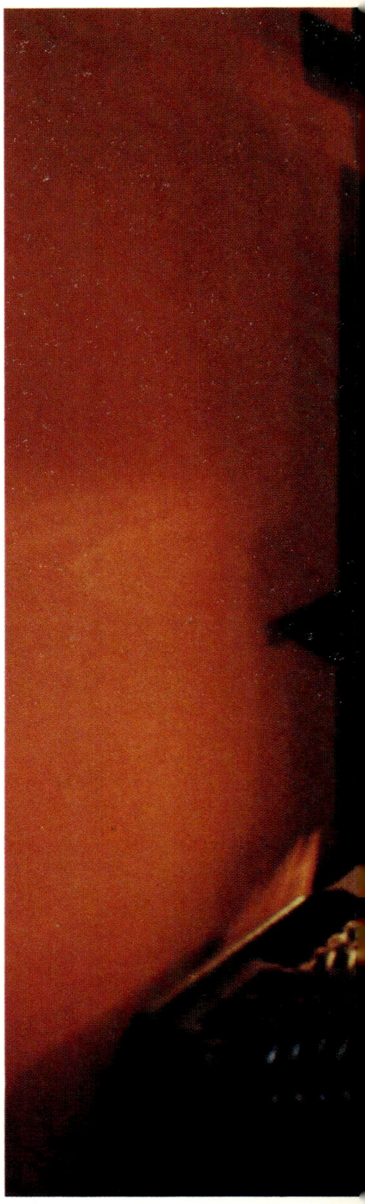

INNOVATION

Swedish prosperity is built on research and development together with a high industrial competence. Almost all Swedish multi-national companies are founded on own inventions. Well-known examples are SKF, LM Ericsson, and Nitro Nobel. All of these had their start around the turn of the century.

Ever since, a high level of innovation has been a tradition in Swedish industry. This is especially true for the many small technically oriented firms continually working on new products and processes.

Sweden accounts for one to two per cent of the world's annual store of new knowledge. This is a remarkable result when you consider our small population. The great advances on the international market during recent years indicates that Sweden's technical level is high. The Swedish Viking spirit lives on.

The core of a micro-computer is a so-called LSI circuit, which is manufactured from a thin slice of silicon. The picture shows electronic measurement of circuits during the production phase at ASEA-HAFO, successful Swedish company in the field of semi-conductor technology.

With the aid of a system based on an infrared technique and gold-plated reflectors, industrial ovens can be heated to 1,300 degrees Celcius in one second.

NUCLEAR ENERGY AND ENVIRONMENT

Swedish industry has had great success in the nuclear energy field, in which Sweden, like the major powers, has been able to develop its own techniques. Nuclear power as an energy source is of special interest to Sweden because it contains 80 per cent of Europe's known uranium resources. There are currently six operational reactors and six under construction.

The environmental debate today is mainly concerned with the safety aspects of nuclear energy. The so-called conditions law, which demands absolute safety even in the storage of nuclear wastes, has prevented any further expansion of nuclear energy.

Research is now aimed mainly at developing renewable energy sources. Emphasis is mainly on wind power, energy forests (biomass) and utilization of solar energy.

A technical and economic breakthrough for the conversion of solar energy is expected by the early 1980s.

Reactors one and two at Barsebäck.

Several solar heating projects are in their prototype stages.

REVOLUTION IN THE FOREST

Like the giant reptiles of prehistory,dinosaures of old, powerful modern forestry machines eat their way through forest clearings. In a matter of minutes they fell and bark the trees and saw them in pieces. The din of modern machinery has inexorably penetrated the forest primeval.

Much more than half of Sweden's area is covered with woods. Swedes are surrounded by soaring spruce and pines that tend to nurture their intermittent melancholy.

In distant times of unrest, the deep forest provided shelter against foreign intruders. It was the peasants' safest refuge when the national authorities became too aggressive. The Swedish forest provided the necessities of life for many a generation of peasants. With the help of the forest a man could make his cradle, his home, his tools, his household utensils and, ultimately, his coffin – the forest shaped his life. In years of failed harvests, the woods provided the berries, mushrooms, and even bark for bread when things were really wretched.

When lumber and pulp became exportable products toward the end of the 19th century, the forest became known as "Swedish gold." The "gilt" has now faded a bit but, nevertheless, about one-fourth of Sweden's exports are still derived from wood.

In the course of a generation, simple forestry has become a great effective industry. The piercing howl of the power saw replaced the handsaw and the axe. Pesticides were introduced. Then giant processors, such as the one from Hasselfors seen here, and ever more ingenious machinery for thinning out and replanting, made their appearance. Today a lumberjack can fell four times as many trees as he could 20 years ago – and his accident rate has diminished considerably.

Many of those who observe this development are worried, and protest against the use of pesticides, gaping clearings, and the machine age. Technology has come too fast to "the pauper's garden."

AGRICULTURE

In the course of a single generation Swedish agriculture has changed dramatically. In 20 years more than half the farms have been given up, fields have been merged into larger units or turned into forest land. Machines have taken over the operation. Agriculture today is a highly effective industry. There are already several completely computerized farms.

Only five per cent of the population earns its livelihood from agriculture, yet people are still being furnished with most of the foodstuffs that can be produced at our rather cool and dark latitude.

Development has simply gone too fast. Somewhere there is a limit to what the soil can stand in the way of artificial fertilizers, pesticides and effective machines. Toward the end of the 1960s, when animals began dying of poisoning and lakes became polluted, we were forced to stop and reflect. Development had to be slowed down.

This is giving the small family farmer a reasonable chance to survive on an income that can compare with that of industry. The countryside can be maintained without further depopulation.

The future of our industrialized agriculture depends to a great extent on trends in the price of oil. But farms cannot be made much bigger and we cannot exploit the earth and livestock much more than we do already. The agriculture of the future will remain much the same as it is today.

If, that is, the price of oil does not double – because then we will have to retrace our steps to the days of the self-sustaining household. Then the small farmer will have his renaissance.

Five per cent of the population lives on farming. But don't believe that this is a quiet minority. The farmers' organizations and farmers' cooperatives are a strong segment of society, and the farmer's influence in Swedish politics is certainly a good deal more than five per cent.

Even in this regard, Swedish agriculture is very efficient.

The most modern technology has now taken over some of the work of the farmer—without making his surroundings less human. Axelvolds Gård in the province of Skåne is one of the pioneers, with 200 cows tended by two men. Computers control the milking, and electronic memory units regulate feeding. But men are still needed to operate the tractors.

Varnhems Church in Västergötland houses 700 years worth of memories.

The debate about women ministers has been raging hot and heavy for years.
Nevertheless, more and more are being ordained in the State Church.

NORDSJÖ

It's becoming all the more natural for women to participate in traditionally male-dominated occupations. No longer is it uncommon to see a woman painter, bus driver or engineer.

Carl Eldh's forceful statue of August Strindberg, one of the few Swedish writers to have established a lasting international reputation.

Nils Ferlin, widely loved poet and song writer, immortalized as a statue on a park bench in his native Filipstad.

The Old Town, as the name implies, is the oldest part of Stockholm and its dominating structure is the Royal Palace at the right.
(preceding pages)

The Masthugg Church stands as a sentry at the entrance to Gothenburg's harbor.

Gothenburg is a sports center as well as Scandinavia's biggest seaport. The new Ullevi Stadium is a compelling part of the city scene.

SPORTS

Sweden is full of organizations as well as of people who like to watch or indulge in sports. It's not surprising, therefore, that two and a half million Swedes—nearly one third of the total population—belong to some kind of sports association.

Sports is the country's largest folk movement. It is unique in that it is so broadly based and cuts across all age and social groups. Once only a privileged few could afford the time and money to indulge in sports. Now the motto is "Sports for Everybody," and we are well on the way toward making this true. Sports has more and more come to mirror society itself. More and more women are participating; there are, for example, soccer leagues for women, and their male companions have ceased to snicker at them—at least out loud.

The old Vikings, who made life miserable for the outside world around 1,000 years ago, are said to have enjoyed competitive swimming and ice skating in their leisure time. (Fencing and boxing were part of their work.) Moreover, the oldest existing sports society in the world is believed to be the Uppsala Swimming Society, dating from 1796.

Soccer is Sweden's most popular team sport in the summer and ice hockey in the winter. While Sweden's world-ranking athletes are comparatively few, they are no less illustrious. Take, for example, Ingemar Stenmark, world champion in slalom skiing, and Björn Borg, triple Wimbledon champion and perhaps the greatest tennis player in the world today. These two stars alone have drawn tens of thousands of enthusiastic youngsters to the nation's ski slopes and tennis courts.

Sweden's King Carl XVI Gustaf and Queen Silvia on a ski holiday in northern Sweden.

Ingemar Stenmark from Tärnaby—world champion slalom skier.

VM
ABC
VM
VM
ABC
JOFA

Ladies' soccer is being taken more and more seriously in Sweden – not only by the participants.

During winter, ice hockey is the big team sport and every youngster dreams of someday making the national team.

Björn Borg, the world's top tennis player.

1209
1034
1379

Trotting is strongly entrenched in Sweden. It provides a greater annual turnover than soccer.

THE ENVIRONMENT

In certain respects Swedes are spoiled. On the strength of the uniquely Scandinavian Allemansrätt, they have free legal access to open property, regardless of ownership, provided they remain out of sight and earshot of a person's private home and do not disturb the tranquillity of his surroundings. Embedded in the soul of all Swedes is a trace of the farmer, the lumberjack or the fisherman. This is a legacy passed on from one generation to another. As soon as he gets the opportunity, a Swede must return to nature—sometimes in a cloud of automobile exhaust fumes. It's been rather easy to arouse public opinion here on the matter of environmental protection. It's everybody's concern.

Foreigners who behold this tranquil land may find it difficult to understand that Sweden has appreciable environmental problems. Much has been destroyed by a rapidly growing Swedish industry, by the automobile and by Swedish thoughtlessness. Thousands of lakes are rapidly becoming acidified—perhaps even the land and forests—and 80 % of the damage is being done by smoke from other industrial nations that are polluting the air. There are no national boundaries when it comes to environmental care.

A lot has been done in the short time we have been aware of the problem. More rigid regulations on the use of gasoline and heating fuels have been introduced, a number of pesticides have been banned for use in agriculture and forestry, and millions of kronor have been allocated for the purification of municipal sewage systems. Where Lake Mälaren meets the Baltic Sea in the heart of Stockholm, the water was heavily polluted only a few years ago. Now, with a little luck, you can even catch a sea trout there—and eat it!

Boats of all types are important and natural parts of the scene along Sweden's coast and on its 100,000 lakes. But, for sailors, the only boat that matters is the sailboat.

It is now possible to fish good healthy sea trout right in the heart of Stockholm. This is a testimony to effective environmental planning.

UCLA

30
SVERIGE
BLÅMES
PARUS CAERULEUS
SVERIGE
FYRA SKILL. B:CO
1924
SVERIGE
ILO 50 ÅR
SVERIGE 55
SVERIGE
SVERIGE
1 KR
1 KR
SVERIGE
FRIMÄRKE
12
TOLF ÖRE
SVERIGE
75
SVERIGE
LENNART HELJE
20 ÖRE
SVERIGE
GUSTAF II ADOLF 6 NOV·1632
10 ÖRE
SVERIGE
CZESLAW SLANIA
1636 20·2 1936
SVERIGE
KUNGL·POST

SVERIGE
5 KR
FRIMÄRKE
FEM KRONOR
EUROPA
ERIC HALLSTRÖM
1.40
SVERIGE
1977
SVERIGE
LUFTPOST
10 öre
1 KR SVERIGE
19·6 1976
CZESLAW SLANIA
SVERIGE
KUNGL
35 ÖRE
FRIMÄRKE
5
FEM ÖRE
SVERIGE
ANDERS SPARRMAN 1748-1820
SVERIGE
10 KR
ERNST JOSEPHSON
CZESLAW SLANIA
SVERIGE
1855-1955
SVERIGE
FRIMÄRKE
4
ÖRE
DROTTNINGHOLMS TEATER 200 ÅR
SVERIGE 5

EUROPE'S LAST WILDERNESS

It may come as a surprise that such a peaceful and well-ordered place as Scandinavia contains Europe's largest—indeed, last—wilderness.

In Sweden alone, there are three National Parks—Sarek, Padjelanta and Stora Sjöfallet—that cover more than 5,000 square kilometers of difficult-to-penetrate terrain north of the Arctic Circle. Soon the far north of Sweden will provide the last chance to experience what this once-wild continent was like for our forefathers.

There you can still see rare flora and perhaps even a bear, wolverine or lynx. There, too, are birds not generally encountered in proximity to man. Strange, isn't it, that our civilized continent has progressed so far that the wild and unspoiled have become measures worth protecting? Why not pick up some heavy shoes, a map and a compass and head up north where you can really envy us?

A powerful part of "Europe's Last Wilderness"—the grand rapids in the Pite River in northern Sweden.
(preceding pages)

Two young Tengmalm's owls and a capercaillie in display are privileged sights for the visitor to northern Sweden's wilderness.

There are close to 600 bears in northern Sweden.

a touch of **Sweden**

SWEDEN	
Constitution	Constitutional monarchy
Head of state	King Carl XVI Gustaf, b. 1946
Successor	Prince Bertil, b. 1912
Parliament	One House, 349 members (from 1977)
Population	8.3 mill.

Distribution of Swedish branches of industry and commerce among different types of owner

Number of employees, %

	State	Municipal	private
All companies	4.4	2.2	93.4
Agriculture, forestry, hunting and fishing	0.3	0.4	99.3
Manufacturing industries	3.5	0.1	96.4
Building industry	0.3	0.0	99.7
General merchandise, restaurant and hotels trades	5.3	0.7	94.0
Communications	12.1	10.7	78.4
Property management and commissioned activities	3.7	17.9	78.4
Mines and quarries	68.0	0.4	31.7
Electricity, gas, heating and waterworks	3.4	47.9	48.8
Services	3.2	3.6	93.1

Population

Largest cities

Stockholm	658,743
Gothenburg	440,211
Malmö	238,567
Uppsala	141,434
Norrköping	120,669
Västerås	118,131
Örebro	117,057

Political parties and election results

	1968 %	1970 %	1973 %	1976 %
Social Democrats	50.1	45.3	43.6	43.6
Center	16.1	19.9	25.1	24.1
Liberals	15.0	16.2	9.4	11.2
Conservatives	13.9	11.5	14.3	15.8
Communists	3.0	4.8	5.3	4.8
Others	1.9	2.3	2.3	3.7

Aliens

Per cent of entire population

1945	0.5 %
1950	1.8
1960	2.5
1970	5.1
1975	4.9
1976	5.2
1977	5.1

Education

Elementary education. A 9-year compulsory comprehensive school starting at the age of seven. English is compulsory from the third year, French or German offered as a second language. Certain options from the seventh year. *The "Gymnasium"*, corresponding to Senior High School in the United States, covers 2–4 years. It offers education and training on both theoretical and practical lines, and is a merger of the traditional Central European "gymnasium" and the former vocational schools.

Universities have existed in Uppsala, Lund, Gothenburg, and Stockholm for a long time, but there is also one at Umeå, and a number of colleges on graduate level throughout the country. The number of students has grown rapidly–from 50,000 in 1963 to 108,500 in 1975.

Adult educations has great traditions in Sweden, and courses are offered along every possible line, nowadays also through TV. Many adult education schemes are offered by organisations like the Labor movement, the Temperance movement, churches, etc.

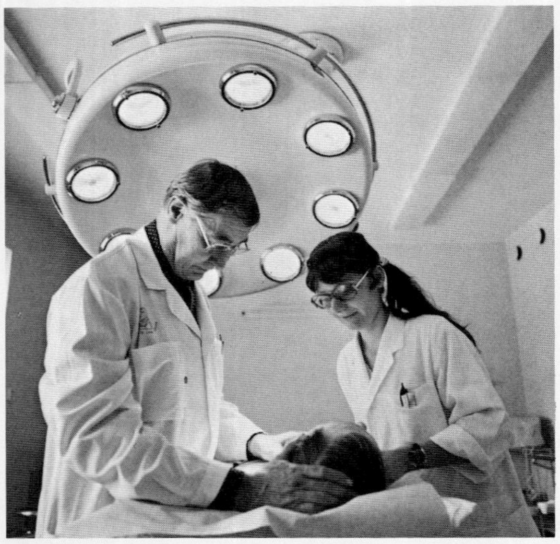

Medicine

Sickness benefits vary from 11 to 218 SKr a day, according to income, and are subject to taxation.

Doctor's fees are refunded in full or in part, as well as expences for transport in many cases.

Prescribed medicines: maximum 20 SKr.

Industrial Injuries Insurance covers all employees, and includes a daily sum, payments for disability, and compensation to dependants in case of death. For *childbirht* every mother is guaranteed 240 days' leave of absence, an outright premium, free dental care, and a daily allowance according to income. The father is permitted to replace the mother, with compensation equal to sick benefits.

Hospitals are administered primarily by the county councils and are of a very high standard. Medical care is freee, out-patients visiting the clinic pay 20 SKr.

Free medical examination is available to pregnantwomen, mothers and infants, and to pre-school children. (Regular check-ups are compulsory in schools.

Dental care is free for school-children, and available at a reduced charge to adults.

Glasses are subsidized for children and young people up to the age of 19.

Old Age

Old Age Pension (basic pension) begins normally at the age of 65 in certain cases at 63. A single person receives about 10,165 SKr, couples nearly 17,000 SKr per year.

Advance Retirement benefits are paid to those who lose their income because of illness or injury.

Widows' Pensions are drawn from the basic pension fund as well as the supplementary pension fund (see below). When there are children these benefits provide good financial security.

Blind and other disabled also receive pensions from the basic pension fund.

The supplementary Pension Scheme, similar to "Social Security" in the United States, was introduced in 1960. Pensions begin at the age of 65 (from 1976).

Sport

Sport is part of the Swedish way of life, and the state-subsidized Swedish Sports Federation has more than 1.5 mill. members.

Soccer and ice hockey dominate among the ball games, followed by handball, basketball, and bandy, similar to land hockey but played on ice. Athletics (track and field) is widely spread–there are 3,700 athletics fields in all. Sweden has more golf courses than any other European country outside the British Isles.

Gymnastics has been a compulsory subject in schools for generations, and a series of "keep fit" programs on TV activated over half a million people!

"Vasaloppet", an annual 53-mile ski race, is the biggest in the world–around 10,000 people usually take part.

Foreign Trade

Exports	%	*Imports*	%
Machinery and transport equipment	43.6	Machinery and transport equipment	31.0
Paper, cardboard etc.	9.0	Minerale oils	16.5
Automobiles, lorries, buses and motor vehicle parts	10.5	Chemicals	8.3
Iron and steel	7.3	Clothes	4.3
Pulp and waste paper	5.8	Food products and live animals	7.6
Iron ore	1.6	Iron and steel	4.1
Food products	2.0	Automobiles, lorries, buses and motor vehicle parts	7.0
Textile fibres	5.8	Raw materials (inedible) except fuels	4.4
Other goods	14.4	Other goods	16.8

Major trade partners	%		%
Export: Great Britain	11.3	Import: West Germany	18.9
Norway	11.2	Great Britain	10.5
West Germany	9.9	East Europe	
Denmark	9.8	(and the Soviet Union)	9.8
East Europe		USA and Canada	7.3
(and the Soviet Union)	9.1	Denmark	7.0
Finland	6.4	Norge	6.1
USA and Canada	5.9	Finland	5.7
France	5.5		

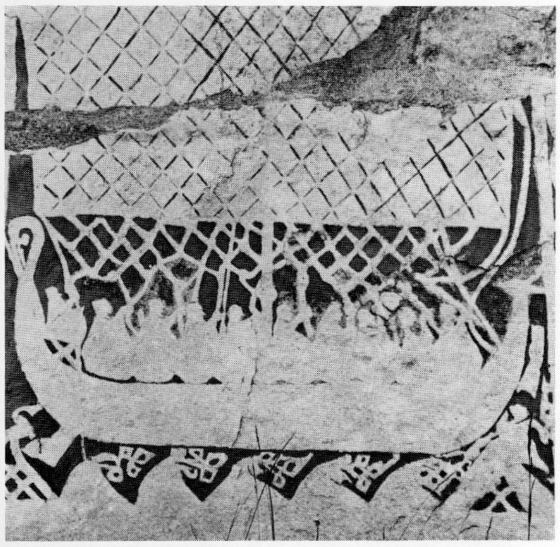

History

500–600	The Svear emerge as the most powerful tribe.
800–1050	*The Viking period.* Swedish Vikings control the seaways to Constantinopel and make trade expeditions to the Orient and Western Europe.
830	The first Christian church founded by Ansgar, a Frankish monk.
1050–1523	*The Middle Ages.*
1164	The first archbishop takes office.
13th cent.	Finland conquered in crusades. New provincial laws written, unique in Europe in age and content.
1397	Nordic Union formed under Danish leadership to fight Hanseatic League.
1521	Swedish rebellion against Danish superiority ends in victory under Gustav Eriksson Vasa.
1523–1721	*The Vasa period. Sweden as a world power.*
1523	Gustavus Vasa crowned. Builds up an administration, and restores the economy by confiscating Catholic Church property.
1611	Gustavus Adolphus, his grandson, ascends the throne. Promotes municipal self-government, science, education, mining, and the arms industry. Further expansion east of the Baltic.
1631	Gustavus Adolphus defeats the Catholic Confederation at Breitenfelt, dies at Lützen the following year.
1648	End of the Thirty Years' war. The Treaty of Westphalia brings Sweden large areas of northern Germany and makes her a world power, supported by France.
1654	Queen Christina abdicates and is converted to Catholicism.
1654–1697	Further expansion and increasing military power but the economy under great strain. Charles XII king in 1697 at the age of 15.
1700–1721	The Great Nordic war begins when Denmark, Saxony-Poland, and Russia under Peter The Great unite to defeat Sweden. Charles the Twelfth killed in 1718. End of the Swedish empire
1721–1818	*A period of domestic expansion*
1723–1772	Constitutional government and a strong parliament, in which four estates (nobility, clergy, burghers, and peasants) are represented. Two political parties emerge. Freedom of the press 1766. Progress in science, agriculture and industry.
1772–1792	Gustavus III makes bloodless revolution to restore royal power in 1772, encourages culture. Assassinated during a masked ball in 1792.
1803	Beginning of land reform aiming at integration of split farm holdings.
1808–1809	War with Russia puts an end to Swedish supremacy over Finland.
1809	New constitution adopted.
1813	Alliance with Great Britain. Swedish troops help defeat Napoleon at Leipzig under Jean Baptiste Bernadotte, former French general.
1818–1918	*End of wars, Modern era begins.*
1818	Bernadotte becomes king of Sweden and Norway. Neutrality is the guiding principle of Swedish foreign policy.
1840's	Decade of liberal opposition among new class of merchants, factory owners, etc.
1842	Elementary schooling made compulsory.
1856	The first railroad opened.
1858	The ban private religious meetings dissolved.
1866	Two-chamber parliament.
1850–1930	More than a million Swedes emigrate to North America.
1880's	Trade unions formed. Temperance movement, feminist movement, beginning of co-ops.
1889	The Social Democratic Party founded.
1905	End of union with Norway.
1914–1918	Sweden neutral during World War I.
1918	*Democracy and non-alignment.*
1918–1921	Universal suffrage for both men and women.
1920	The first Labor administration.
1921	Capital punishment abolished.
1939–1945	Sweden stays neutral in World War II.
1948	Sweden decides not to join NATO.
1955	Compulsory health insurance.
1957	Social security program (ATP) adopted after referendum.
1959	Sweden becomes member of EFTA.
1970	One-chamber Parliament.
1971	Sweden decides not to apply for full Common Market membership.
1975	The King's functions reduced to ceremonial tasks only.
1976	After 44 years, the Social Democrats had to hand the ruling of Sweden to a Government consisting of three non-socialist parties.

Industry

Largest export industries

Company	Main products	Total sales abroad (mill. SKr)	Number of employees
AB Volvo	Automobiles, buses, tractors, trucks, dumpers, jet engines, farming and forestry equipment	15,742.9	62,755
SKF	Bearings	6,981.5	58,041
LM Ericsson	Telephones, telecommunications equipment	7,311.5	71,100
Statsföretag AB	State-owned group of companies. Sales abroad dominated by iron ore	9,747.3	47,710
AB Electrolux	Vacuum cleaners, refrigerators, freezers, office machines	7,645.7	71,552
Johnson Group	Shipping, iron and steel, oil distribution	12,078.7	29,434
ASEA	Electric equipment and machinery, power plants (also nuclear)	8,400.2	44,194
Svenska Tändsticks AB	Building materials, matches	4,622.0	33,850
Sandvik AB	Special steels, alloys	3,908.6	27,891
SAAB-Scania AB	Automobiles, trucks, airplanes, electronics	9,613.0	41,386
Alfa-Laval AB	Separators, thermal equipment	3,776.6	17,820
Atlas Copco AB	Compressors, rock drilling equipment, industrial tools	3,791.2	18,384
Gränges AB	Mining, iron and steel, metal manufacturing	5,379,1	24,630

The Budget

Central Government expenditure

Health and social welfare	23.3 %
Education	12.7
Defense	9.5
Labor market and regional development policy	5.3
Tax equalization grants to municipalities	4.2
Transport (roads and bridges 2.6 %)	4.2
Agriculture and forestry	4.7
Law courts, police, prisons, etc.	3.4
Other expenditure	19.3

	Gross income:	Total tax:
Income tax for single persons and married persons (if both man and wife have an income)	SKr 30,000	30 %
	50,000	50
	80,000	60
	100,000	65
	150,000	70
	200,000	75

Total taxation
(per cent of gross national product)

Sweden	43.1 %
Norway	46.1
Denmark	45.7
Netherlands	45.8
West Germany	37.5
France	35.9
Great Britain	35.7
Italy	30.9
USA	29.3
Japan	23.0

Transportation

Registered automobiles (passenger cars)

Volvo	808,356
Saab	387,369
Volkswagen	385,657
Opel	304,634
Ford	240,592
Mercedes-Benz	121,474
Renault	99,006
Peugeot	89,241
Fiat	87,315
Other makes	32,324

Standard of living

Housing, etc.

Piped water	98 % of all dwellings
Central heating	96
WC	94
Private bath or shower	87
Freezer	77.3
Dish washer	17.6

Cars per 1,000 inhabitants=3.1 inhabitants per car	343
Telephones per 1,000 inhabitants	633
TV licences per 1,000 inhabitants	369
Charter flight passengers to foreign countries	908,808
Paper consumption (per inhabitant)	220 kilos
Consumption (per inhabitant) of deep-frozen food, excluding icecream	23 kilos

The Nobel Prize

When Alfred Nobel, industrialist and inventor of dynamite (1833–96), made his will in 1895 he founded a prize of great international prestige, now amounting to 681,000 SKr. Prize-winners for Physics, Chemistry, Physiology or Medicine, Literature and Economics (since 1969) receive their awards in Stockholm, while the prize for Peace is presented in Oslo.

The names of many of these men and women are written in the history of our century–Henry Dunant, President Wilson, George Marshall, Winston Churchill, Albert Schweitzer, Bertrand Russell, Thomas Mann, George Bernard Shaw, Alexander Fleming, Linus Pauling, Marie Curie, Albert Einstein, Enrico Fermi, Willy Brandt, Henry Kissinger.

Nobel Prizes, most represented countries

USA	134	Belgium	13
Great Britain	72	Italy	11
Germany	59	Denmark	11
France	41	Netherlands	10
Sweden	23	Austria	10
USSR	15	Norway	7
Switzerland	13	Spain	1

Religion

The Church of Sweden

was firmly established as an Evangelical–Lutheran church in 1593. It is still closely tied to the State in terms of organization and economy but is independent in religious matters. All Swedish citizens are members unless they have withdrawn formally or were born to parents who do not belong. Members of independent denominations are frequently members of the Church of Sweden as well. Since 1960 women can be ordained as ministers. *The Roman Catholic Diocese of Stockholm* has 81,022 members.

Independent denominations (membership)

Pentecost Movement	94,970	Swedish Baptist Church	22,898
Swedish Missionary Society	82,314	Örebro Missionary Society	19,684
Salvation Army	36,310	Swedish Alliance Missionary Society	13,393
Evangelical National Missionary Society	25,589	Jehovah's Witnesses	12,151
		Methodist Church	6,118

Technology and Research

This radio telescope at the Råö observatory near Gothenburg symbolizes the fact that the wealth of Sweden depends largely on technological and scientific skill–past, present and future. Major questions pertaining to scientific research are handled at cabinet level by the Research Council, with the Prime Minister as chairman. Commercial and industrial interests dominate government-sponsored projects, however: by 60 % to 40 %, in terms of money.

Some Swedish inventions

Safety matches	Gustaf Erik Pasch	1844
Screw propeller	John Ericson (and Francis Smith)	1836
Dynamite	Alfred Nobel	1867
Steam turbine	Gustaf de Laval	1889
Milking machine	Gustaf de Laval	1896
Automatic sun valve (for lighthouses)	Gustaf Dalén	1906
Spheric ball bearing	Sven Wingquist	1907
Refrigerator (absorption method)	Baltzar von Platen–Carl Munthers	1922

Flying Time: Stockholm→

Gothenburg	50 minutes
Malmö	1 hour 5 minutes
Kiruna	1 hour 35 minutes
Luleå	1 hour 10 minutes
Rome	4 hours 30 minutes
Athens	5 hours 20 minutes
Moscow	2 hours 10 minutes
Madrid	5 hours 25 minutes
Frankfurt	1 hour 55 minutes
Berlin	2 hours 55 minutes
Peking	14 hours 50 minutes
London	2 hours 30 minutes
Paris	2 hours 30 minutes
Copenhagen	1 hour 10 minutes
San Francisco	16 hours 45 minutes
Seattle	12 hours 20 minutes
New York	10 hours 35 minutes
Los Angeles	13 hours 35 minutes
Chicago	11 hours
Montreal	10 hours 5 minutes

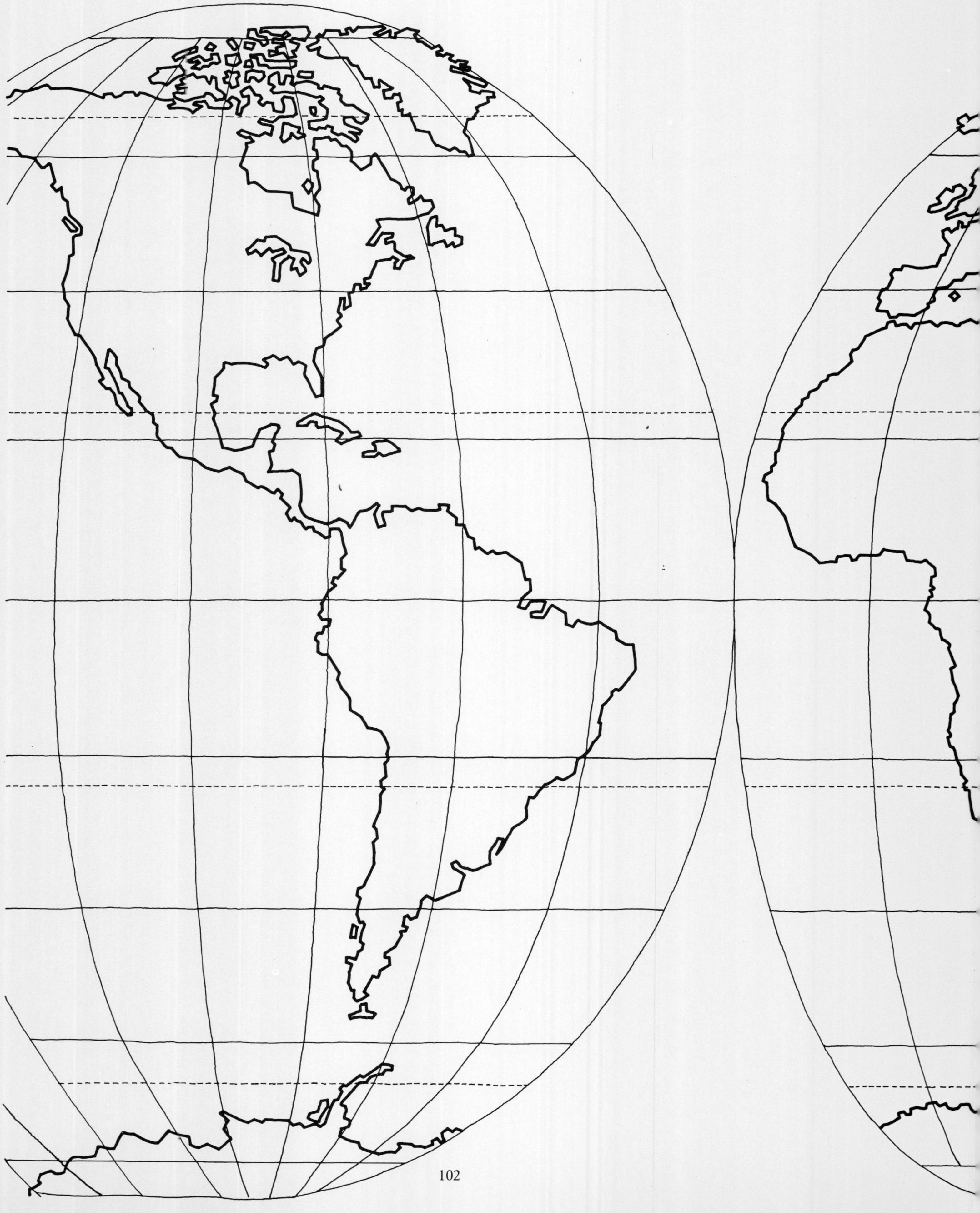

Toronto	**11 hours 35 minutes**	**Tokyo**	**15 hours 30 minutes**	**Colombo**	**15 hours 55 minutes**
Vancouver	**14 hours 50 minutes**	**Hong Kong**	**19 hours 50 minutes**	**Hawaii**	**26 hours 25 minutes**
Mexico City	**17 hours 25 minutes**	**Singapore**	**16 hours 40 minutes**	**Cairo**	**8 hours 15 minutes**
Caracas	**17 hours 5 minutes**	**Sydney**	**26 hours 20 minutes**	**Cape Town**	**20 hours 5 minutes**
Buenos Aires	**22 hours 35 minutes**	**Melbourne**	**27 hours 30 minutes**	**Addis Abeba**	**11 hours 35 minutes**
Rio de Janeiro	**17 hours 30 minutes**	**Auckland (New Zealand)**	**33 hours 25 minutes**	**Casablanca**	**7 hours 20 minutes**
Sao Paulo	**19 hours 15 minutes**			**Laos (Nigeria)**	**10 hours 30 minutes**

Books by Gullers

1946 Stockholm – The summer city
1947 Sweden – USA, Ziff Davis
1947 Gothenburg
1948 Sweden from the air
1949 Italien
1949 Stockholm
1950 I kvinnans värld
1951 England – GB – Odhams Press
1951 Norge – N – Tidens Förlag
1952 Teknis i bild
1952 Gullers' New Book about Sweden
1953 The old city of Stockholm
1953 Norrköping
1954 Sverige – Vårt Land
1954 Howard Foundry, USA
1955 Foto för alla: S, N, DK, SF
1956 Uddeholm
1956 Sweden – The Water Land
1957 White Christmas
1957 The Metal Way
1958 Carina with Love, GB – Macmillan
1959 Vikmanshyttan
1959 Those Happy Years
1960 Italien – pocket: S, NL, US
1960 Sweden – a Journey through the Ages
1961 Findus Saga
1962 Mölnlycke
1962 Wifstavarv
1963 Made in Sweden
1964 Gullers Sweden
1965 Sweden – pocket
1966 Working together in Sweden
1966 Sieverts Kabelverk
1967 A Bit of Sweden
1968 Sweden around the World
1968 Stora Kopparberg
1968 USA – Sweden
1969 Knäpp
1970 Sweden your new Country
1970 France – Suède
1971 Sweden – pocket
1972 ICA 50
1972 The Face of Sweden
1973 ASEA
1973 Götabanken
1973 The Crewel Needlepoint World
1973 Fit for Fun
1974 Då
1974 Handarbete på nytt sätt
1975 Boliden
1975 The Face of Sweden – pocket
1976 Sea Songs
1976 SL 100
1976 Uppsala University 500
1977 Swedish Know-How
1977 Linnaeus
1977 Dalsland
1978 Pripps 150
1978 A touch of Sweden
1978 Esso in Sweden

Photography: K W and Peter Gullers, Björn Enström, Einar Anderssons Pressbild, Ö. Björkdal, Pressens Bild, Hans T. Dahlskog, Pressens Bild, Åke W. Engman/N, Boccon Gibod, Svenskt Pressfoto, Leif R. Jansson, Svenskt Pressfoto, Peter Lindberg/N, Peter Knopp, Reportagebild, Olle Kymling, Östen Matsson, Bo Möllstam, Reportagebild, Bertil Pettersson/N
Camera: Hasselblad, Leica
Helicopter pilot: Stig Ekdahl
Reproduction: Litoteam AB, Gothenburg, Sweden
Printing: Ljungföretagen, Örebro, Sweden
Adaption: Richard J. Litell, Stockholm, Sweden
Publisher: Gullers International ab, Stockholm, Sweden
ISBN 91-85228-33-8